All Long Island from the air, 1962. Largest island off our Atlantic seaboard, Long Island has an area of 1,682 square miles, a width of 23 miles, and shore lines totalling about 500 miles. The coast has intimate clarity in this air photograph, which reveals also indications of glacial moraines, plant cover, and the hundred or more square miles of salt-marsh. (Photograph used by courtesy of Lockwood, Kessler and Bartlett, Inc., Consulting Engineers, Syosset, N. Y.)

LONG ISLAND

Continuous lines mark the terminal positions of the two latest advances of the Wisconsin glaciation. That passing near the north shore through Orient Point is the Harbor Hill moraine; the other, running through Montauk Point, the Ronkonkoma moraine. Highest elevations above present sea level, in feet, are at Harbor Hill and Manetto Hills.

The treeless Hempstead Plains are as drawn in the Robert Ryder survey, *circa* 1675.

Territory occupied in the early seventeenth century by the principal Indian communities or "tribes" is indicated by thirteen names in small capitals, several of which appear on the map twice or thrice: Canarsie; Merrick; Rockaway; Massapequa; Secatogue; Patchogue; Shinnecock; Montauk (the peninsula and Gardiner's Island); Manhasset (Shelter Island); Cutchogue; Setauket; Nissequogue; Matinecock.

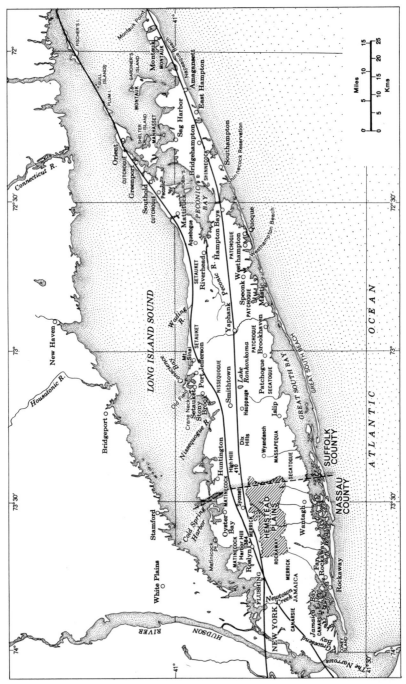

Geography and communities of Long Island. (For explanation, see the reverse of this page.)

FISH-SHAPE PAUMANOK

Nature and Man on Long Island

ROBERT CUSHMAN MURPHY

American Museum of Natural History

Penrose Memorial Lecture for 1962

THE AMERICAN PHILOSOPHICAL SOCIETY

INDEPENDENCE SQUARE · PHILADELPHIA

1964

This book is illustrated by a map, reproductions of paintings, photographs, and by sketches made between 1833 and 1867 by William Sidney Mount (1807-1868), selected from the Melville Collection in the Suffolk Museum of Stony Brook, Long Island.

Preface

The Penrose Memorial Lecture for 1962, by Robert Cushman Murphy, Lamont Curator Emeritus of Birds at the American Museum of Natural History, is the twenty-eighth in a series established in 1934 by the American Philosophical Society in recognition of a large bequest from one of its members, Dr. Richard A. F. Penrose, Jr. The list of Penrose lecturers, members or guests of the Society, now includes many well-known scientists, scholars and men of affairs.

Surely, however, none of these eminent men spoke with greater affection for his subject, nor clothed his learning in more colorful language, than Dr. Murphy, when he told the story of nature and man on Long Island, the land of his own birth and his lifelong home. As a boy he learned to know the varied terrain of the island, from the Sound to the sea, from Montauk to the Narrows, and the creatures, plants, animals and men that lived upon its soil and its sands or swam in its waters. In later years, as a naturalist of far-ranging experience, he views these scenes of his youth with deeper understanding, through the eyes of a scientist and student of human history.

Likewise a native of Long Island, Walt Whitman has written of "fish-shape Paumánok" with a poet's abandon. Dr. Murphy, choosing some of Whitman's lines as an introduction to his lecture, gave, in prose worthy to follow the poet's verse, a scientist's precise account of the land and its inhabitants, tinged with nostalgia for bygone days and vanished creatures, yet filled with a joyful sense of the timeless beauty of earth and living things.

G.W.C.

CONTENTS

FISH-SHAPE PAUMANOK

Fish-Shape Paumanok

Starting from fish-shape Paumanok where I was born,
Well-begotten, and rais'd by a perfect mother —
O to go back to the place where I was born,
To hear the birds sing once more,
To ramble about the house and barn and over the fields once more,
And through the orchard and along the old lanes once more.
O to have been brought up on bays, lagoons, creeks, or along the coast,
To continue and be employ'd there all my life,
The briny and damp smell, the shore, the salt weeds exposed at low
     water,
The work of fishermen, the work of the eel-fisher and clam-fisher.
O the sweetness of the Fifth-month morning upon the water as I row
     just before sunrise toward the buoys,
I pull the wicker pots up slantingly, the dark green lobsters are des-
     perate with their claws as I take them out.
Sea of stretch'd ground-swells,
Sea breathing broad and convulsive breaths,
Sea of the brine of life and of unshovell'd yet always-ready graves,
Howler and scooper of storms, capricious and dainty sea,
I am integral with you, I too am of one phase and of all phases.
I too Paumanok,
I too have bubbled up, floated the measureless float,
     and been wash'd on your shores,
I too am but a trail of drift and debris,
I too leave little wrecks upon you, you fish-shaped island.

Thus Walt Whitman, with his lifelong nostalgia for
Long Island and his Quaker rendering of the month of
May, answers the question of my friends who were puzzled
by the title of this Penrose Lecture. If they prefer, I might
call it "the history of a detached fragment of Atlantic
coastal plain during the period between the retreat of the
Wisconsin ice sheet and the building of Levittown!"

1

June 7th 1849 —
= Clam diggers waiting for the Tide —

Red shirt & Yellow
Brown basket.

Clammers and fish-spearers.

2

Credentials from a home locality are indispensable for anybody who dares talk aloud about Long Island or the highly proprietary old-time Long Islanders. I recall from boyhood an obituary tribute to a country physician, which guardedly stated: "Although not a native, Dr. Gildersleeve lived in our community for 49 years and had endeared himself to many residents." Therefore, I should say that on the paternal side I am a fourth-generation inhabitant of fish-shape Paumanok. My personal identification with

Glacial erratic boulders.

the area is attested by accounts in my bound journals, which are now in the Library of the American Philosophical Society, of four successive emergences of the seventeen-year cicada, Brood 14. These occurred in 1906, 1923, 1940, and 1957. Undoubtedly, I also saw—or heard—the noisy manifestation of 1889, at which date I overlooked the possibility that our Society might some day consider it as promoting "useful knowledge." I am now looking forward to 1974 and a sixth appointment with the homopterous Rip van Winkles.

3

In the Beginning

Long Island was both created and made lifeless by the latest Pleistocene glaciation, which left only hard-part remains of organisms from preceding times. After the ice of the Wisconsin advance had gradually melted northward, the exposures were stark sand and glacial till and erratic boulders, some of the last as big as Philosophical Hall. These rocks had been given an inexorable ride so directly down the meridians that, thousands of years later, they were correlated from west to east, with the skeletons of the Adirondack, Laurentian, Green, and White mountains. Credit for this has sometimes been assigned to the elder Reverend Timothy Dwight, who was not primarily a geologist. But a re-reading of the shrewd observations made by this scholarly divine during his travels on horseback throughout Long Island shows that the Biblical Deluge was closer to his mind than the Wisconsin glaciation. Beneath the Pleistocene layers are the Cretaceous and pre-Cambrian unconformities, of which only the former has any part in modern landscape and the reservoirs of ground water.

Terminal moraines were left by the ice cap as two ranges of axial hills. These, mostly less than 350 feet in altitude, are the island backbone today. High Hill, south of Huntington, culminates in a sharp knob 410 feet above sea level, the greatest elevation in its latitude to the east of the New Jersey and Pennsylvania mountains. Eskers were deposited along the valleys of streams flowing from the watershed toward Long Island Sound or the Atlantic. From the ice foot, where the glacial front had finally rested, the drainage ran more gently southward across the outwash plain that ended at an Atlantic raised to new high levels by the melting of icecaps. Underneath, lenses of sand, squeezed within Cretaceous clay, contained the cool, aerated, and abundant water of Long Island springs. Some of these aquifers were of the usual kind, replenished by rainfall from the surface, but others were sealed pockets of non-renewable or "fossil" water, which for long held a strong artesian head.

Spring-fed mere above the salt marsh of Flax Pond, Old Field. A half-submerged glacial boulder is in the middle distance.

The withdrawing ice also left behind countless laggards of its own substance. Blocks of the glacier that plowed deeply into the substrate had become buried and insulated by aggradation. These persisted, in some instances for many centuries until, after melting, they left kettle holes great or small. One called Whitman's Hollow, on the Huntington property of one of Walt's ancestors, appeared on a map as early as 1694. The little kettles, alas, are today favorite receptacles for spoil and rubbish. Some have drained dry, others keep marshy bottoms, and still others remain as ponds. The biggest of them all, Lake Ronkonkoma, is 55 feet above sea level, 4,300 feet in long diameter and 97 feet in maximum depth.

Flood Tide of Life

Now, as ice waned and the climate of our current inter-glacial epoch warmed, naked Long Island lay ready for reinvasion by land and sea and air. Life returned across seasonally frozen straits, afloat on ocean drift, through currents of the atmosphere, and by way of the alimentary tracts of mobile animals. The early arrivals would have been mostly from southerly quarters but later repopulation came from all points of the compass. Each of the physiographic phenomena already mentioned had a part in selecting and controlling the reestablishment of life. The truth of this

Port Jefferson.

7

applies not only to the first reoccupation by humble plants and scavenging arthropods, but equally to the later distribution and ecology of both savage or preliterate man, who was in essential balance with the environment, and of seventeenth- to twentieth-century man, who was and still is vastly more of a meddler and upsetter.

The sources for new life were peculiarly rich in our western hemisphere, a fact which is only an additional result of glaciation. In Eurasia the ice sheet pressed southward toward transverse mountains—the Pyrenees, Alps, Caucasus, Altai, and Himalayas. A wide variety of organisms, including most of the native northern trees, was wiped out beyond hope of revival. In North America, on the contrary, the Appalachian and Rocky Mountain ranges extend prevailingly north-south. The forests were pushed back to a latitude which was not lethal, and were able to recapture their old territory in the wake of the retreating ice. That is why Europeans found in eastern temperate North America four times as many kinds of trees as survived in all postglacial Europe and Asia to the north of the mountains.

Climate and Rhythms

Long Island has almost unique climatic characteristics. In the first place, total annual wind transport is greater than anywhere else on the Atlantic seaboard. This means that breezes blow more steadily, rather than harder. The mesoclimates vary remarkably, moreover, throughout the length of the island. The unlikeness of the two ends, less than 118 miles apart, is great enough to account for about twelve days' difference in the spring leafing of trees and shrubs. There is progressive diminution of mean air temperature from west to east. The same is true of the ocean and it is, in fact, the influence of cool surface water, rather than of the land, that retards the early growth of vegetation in eastern Long Island. On the other hand, well into autumn and winter, the latent heat of the ocean has such a softening effect upon the climate of the eastern flukes that the growing season, which is the period between killing frosts, is longer there than at the snout of the fish-shape island, close against the mainland. The ocean temperature at the eastern end is actually higher in winter than that around the western end, despite the fact that it is colder in summer and in the average for the year. Thus Montauk, despite its lower mean temperature, has a longer frostless season by fully three weeks than Roslyn, the latter being more affected by the cold northwesterly winter weather of the continental climate.

The delicate controlling effect of the temperature of sea water upon blossoming is graphically shown by two shrubs on our own property at Old Field. These are both *Franklinia,* first found in, and transported from, Georgia

9

by John and William Bartram in the eighteenth century, later to disappear inexplicably from the site of discovery. Our plants are of the same age and size; they grow in common soil and are equally exposed to the sun. But one of them stands approximately 150 yards from mean high-water mark, the second nearly twice that distance. Not only does the latter first put forth leaves in spring but, moreover, each midsummer finds it in full flower when its fellow is only in bud. On the other hand, certain species of plants which grow close to salt water, as well as in the "frost holes" of the inland pine barrens, show a reversed seasonal discrepancy, the seaside examples being the first to bloom.

Trying to keep warm.

Ever since glaciation, Long Island has undoubtedly experienced both secular and short-term climatic rhythms.

10

Plentiful meteorological data substantiate that "grand-father's cold winters" were neither figments nor experiences due to lack of modern house-heating. As we look back a century and more, we know that the winters were, on the average, severer than at present. Most appalling of all weather dates on Long Island fell in 1816, when there was frost in every one of the twelve months. The Fourth of July was celebrated by citizens wearing mittens and greatcoats. This calamitous season was probably not part of a climatic trend. Rather, its phenomena have been attributed to an extraordinary screen of dust in the higher atmosphere from preceding eruptions of Soufrière and Tambora, volcanoes of the West and East Indies, respectively. At any rate, the only corn that ripened to provide next year's seed in Suffolk County, Long Island, grew on Strong's Neck, flanked by the ameliorating waters of Setauket Harbor and Conscience Bay.

Scarecrow in the corn.

Hurricane "Carol," August 30, 1954, at Old Field Point. (Courtesy of the late Colonel Leonard Sullivan.)

My Long Island grandfather, who was born in 1821, grew up among recollections and gossip of "the year without a summer." I have thus far searched in vain to find out whether this was likewise true of Walt Whitman, who was born two years closer to the dire time.

Long Island hurricanes, of which my generation and several following have vivid memories, may be dismissed cavalierly. The widespread impression that they are something new in our world is not to be credited. The paths of devastating hurricanes along the eastern seaboard have been plotted with considerable, even if irregular, frequency for the past hundred years. I learnt this when discovering thirty years ago that finds of tropical seabirds in northern North America invariably follow the passage of such cyclonic storms. Their "newness" is only a correlation with human population and occupancy. When hurricane waves swept across Westhampton Beach before a house had been built throughout its length, the facts were hardly news. But when scores of residences are battered from their foundations, and corpses wash into the bays, the cause gains boundless notoriety. It is simple truth that casualties due to storm and erosion on Long Island result mainly because families persist in founding their habitations in places best reserved for fair-weather picnics. The usual recourse, however, is to blame the Government.

Nature's Communities

Prehistoric Long Island developed a rich variety of habitats for an area of such slight elevation. Nine distinct examples are the following:

a. White and pebbleless ocean strands. Behind these, in interdune swales of barrier beaches facing the Atlantic, was a plant association of special interest which included groves of age-old hollies. In sunnier parts were acres of the yellow-flowered prickly pear *(Opuntia)* and of the beach plums that drew enthusiastic praise from Henry Hudson when he landed at Coney Island in 1609.

b. Vast expanses of salt meadow, bordering or enclosing bays and creeks that furnished food and shelter for

Edge of a white-cedar swamp near the Peconic River, Riverhead.

14

ducks, geese, waders, and other marsh fowl. These tidal meadows were likewise the nursery of shellfish.

c. White-cedar swamps of the once extensive fresh-water wetlands. The famous and eminently useful thick-boled tree, the southern white cedar, has the unique distinction of filling space above ground with a cubic volume of wood greater than that of the empty air between its crowded trunks.

The Stony Brook white oak, which has a breast-high diameter of nine feet. No boring has been taken to determine its age.

d. Inland pine barrens, where the pitch pine was dominant in a special plant and animal community. The undercover was poorer than that of the neighboring continental pine barrens, but it contained many invading plants of the southerly coastal plain as well as heaths from the north. This tract was the stronghold of the vanished heath hen or eastern prairie chicken. Except for Cape Cod, it was and is the only country south of or below the mountains where—nobody yet knows why—the hermit thrush breeds and sings. Possibly, however, old Walt has supplied a worthy ornithological clue, for he "heard at dawn the unrivall'd, the hermit thrush from the swamp-cedars." Might it be that the often parched pine barrens have inherited the matchless songster from wet and green-black groves of *Chamaecyparis* that had once adjoined them?

e. Deciduous timber land of the rolling and glaciated northern strip, where many kinds of hickories and magnificent oaks mingled with about seventy other species of trees. This was the home of the white oak, the delight of shipwrights. One example, now moribund, near my home has a diameter of nine feet at shoulder height. But the groves are gone, from Long Island and the world, except on Gardiner's Island where a primeval stand of 250 acres has been hacked by hurricanes but never by an axe.

Fishing in Long Pond.

16

f. Prairie, such as the Hempstead Plains, on which it was quaintly believed by many inhabitants that trees could not grow. This territory originally supported a distinctive herbaceous flora that has in large part disappeared. There also market shooting for the golden and upland plovers and the now extinct Eskimo curlew was carried out by droves of gunners who walked beside horse-drawn or ox-drawn carts. The beasts calmed wariness out of the game and the wagons were piled high.

g. Colorful downs, such as at rolling Montauk, where the winds restrict tree growth to sheltered hollows.

h. Ponds and fresh-water bogs which are the head-waters of streams. These were the abode of orchises, pitcher-plants, sundews, and cardinal flowers, which still hold out, though far between. But I know one bog where in June I can still see at a glance a hundred thousand nodding blossoms of the snake's-mouth orchid.

Long Island has five rivers and was once a land of innumerable brooks. Most of the latter dried up after the felling of the upland forest. We can now find *extinct* brook-beds with cutbanks, brinks of waterfalls, and potholes— everything indeed save a drop of water. One such old wind-ing course has become, too appropriately, "Death Valley." Walt Whitman called Paumanok "Isle of sweet brooks of drinking-water," yet he himself foretold the thirsting change when going back after years of absence to the farm of the Van Velsors, his mother's family, near Cold Spring Harbor. He noted: "Even the copious old brook and spring seem'd to have mostly dwindled away."

i. The fiordlike harbors along the western half of the Long Island Sound shore. These are, or were, under wood-land shade down to the edge of the shingle beach and the tier of boulders, dry or drowned.

Postglacial Insularity

About 1600, a generation before European settlement, all of these different yet interdigitated types of country swarmed with such an abundance of native life that it is hard even for our imaginations to reconstruct the picture. Daniel Denton, who in 1670 published the first description in English of Long Island, wrote of the wildflowers, which cause "the Countrey itself to send forth such a fragrant smell, that it may be perceived at Sea before they can make the Land."

A great proportion of these indigenous plants have shrunken away to remote hideouts, through man's wantonness rather than necessity. The place in the Long Island landscape of lobelias, ladyslippers, columbine, the most gorgeous of the many milkweeds, gentians, cowslips, wakerobin, lupines, mallows, and many more has been appropriated by such introductions as the daisy, chicory, Scotch thistle, burdock, Queen Anne's lace, Japanese honeysuckle, the most disfiguring of all pests in the eastern United States, and the reed *(Phragmites)*, which is ousting the cattail.

The animal life comprised almost everything that might be expected, and more. The early printed record, though, is scanty. I can find nothing, for example, about beaver-trapping on Long Island and yet beaver teeth turn up whenever archaeologists shake the stuff of middens through their sieves, and the term "beaver dam" appears in written records. Along every waterway in the proper seasons the nights were loud with the clamor of ducks, thirty kinds or more. The sky-darkening flights of passenger pigeons crossed over twice a year. The rivers became seasonally choked with

salmon, sturgeon, and alewives. Among the bear-oak shrub-
bery of the pine barrens the heath hen assembled in dawn
circuses to boom and dance, a performance which the
Indians copied in their own rituals. And who, alive today,
can conceive that a woman walking the Long Island dunes
in 1700 counted thirteen stranded whales between East
Hampton and Bridgehampton, with innumerable others
spouting and breaking water offshore? Denton's account
expands this:

> Upon the South-side of Long-Island in the Winter,
> lie store of Whales and Crampasses, which the inhabitants
> begin with small boats to make a trade Catching to their
> no small benefit. Also an innumerable multitude of Seals,
> which make an excellent oyle: they lie all the Winter upon
> some broken Marshes and Beaches, or bars of sand before-
> mentioned, and might be easily got were there some skilful
> men would undertake it.

The "skilful men" took over all too soon. We still see
harbor seals in icy winters but their principal memorial is

Setauket homestead.

19

in the name of two Robin or Robin's islands, at least one of which was formerly "Robben," the Dutch word for seals.

It is not quite true that *everything* to be expected might be found on Long Island. Its insular status was emphasized by the absence of certain organisms that had never made the seemingly easy skip from the mainland. In addition to the missing pine-barren undercover of New Jersey, Long

Chipmunk.

Island has always lacked the red squirrel, which abounds on the adjacent continent. The copperhead, still a resident of the Palisades of the Hudson, is missing, as are also a widespread toad *(Bufo americanus)*, the wood tortoise, and two small lizards that extend considerably farther northward in New York and New England. It is curious that Long

Island should lack the lizards yet have twelve species of salamanders and nine of frogs and toads, to which sea water would be much more hazardous than to a reptile. A number of neighboring winged animals are also wanting. The tufted titmouse breeds on Staten Island, New York City, yet has rarely if ever crossed a mile over the Narrows to become established on Paumanok.

Overlapping Life Zones

There is perhaps no better way to point out the distinctness of Long Island and its oceanic environment than to contrast them with the Mediterranean Basin in the same latitude. That sea and the outlying parts of the eastern North Atlantic are climatically equable. The fauna is of relatively fixed, static, and temperate type, even though many of the species seem to be of ultimately tropical origin.

On and around Long Island we have, strictly speaking, neither a temperate climate nor a temperate assemblage of life. Rather, under the influence of the Gulf Stream from one side and the Labrador Current from the other, we find an extraordinary mingling of almost subtropical and subarctic conditions, marked by great seasonal shifts. We live under extremes—hot summers, cold winters—circumstances that are quite un-Mediterranean. The wide amplitudes mentioned pertain, however, to the marine environment rather than the land. The range of terrestrial temperature on Long Island is slighter than in most parts of the United States.

Our life might be called half southern, half northern. The holly, persimmon, tupelo, black walnut, southern white cedar, gayfeather *(Liatris)*, cactus, and pink moccasin-flower from the south overlap the heath *Hudsonia,* the bearberry, Labrador crowberry, and red spruce, here at their southernmost outpost. The gray fox, opossum, water snake, box turtle, barn owl, cardinal, prairie warbler, Carolina wren, mockingbird, and pale dune grasshopper in like manner come up the coastal plain to meet such northern species as the trailing arbutus, black-throated green warbler, and

22

hermit thrush. Even the saw-whet owl and the red cross-bill, both birds of the Canadian Zone, have at least once joined the hermit thrush in nesting on the pine barrens. Similarly, at sea the squeteague overlaps the cod; the diamond-backed terrapin (a once-abundant breeder on Long Island) meets the harbor seal; the big warm-water sea turtles, and even a few flying fish, invade cold waters that favor the lobster.

Many southern organisms drop out of the list, one by one or abruptly in groups, as the coastal plain stretches northeastward from the Carolina capes toward Cape Cod. The nearly three hundred species of pine-barren plants in

Eel Spearing. Painted at Setauket, 1845, by William Sidney Mount. The boy has been identified as Tom, one of the sons of Judge Selah B. Strong, the woman as Rachael Holland Hart. Descendants of both still reside in the community. (Courtesy of the New York State Historical Association, Cooperstown.)

southeastern Virginia are successively reduced by each of the waterbreaks, such as Chesapeake, Delaware, and New York bays and Long Island Sound. Only a third of the Virginia species reach Cape Cod and a mere handful the coast of Maine. There is a similar elimination of animals along the same route. The Carolina chickadee stops sharply at the Raritan River in New Jersey; the opossum and the cardinal go on as far as Cape Cod.

Delicacy in the adjustment of a form of life to climatic controls is well shown by an apparent correlation that I have discovered. Let us begin the example with the geographic setting. In only two small and restricted parts of the globe can the annual extremes of temperature of the ocean surface amount to as much as 50°F., or from slightly below the freezing point of fresh water to a maximum of 80° or thereabouts. One is in the western Atlantic, roughly between Cape Hatteras and Cape Cod. The other is to the southeast of Kamchatka, in the corresponding district of the western North Pacific. It may well be significant that a beach herb, the dusty miller *(Artemisia stelleriana)*, grows on the sands of both these oceanfronts which are so remote from one another. The history of the plant is somewhat obscure. It was apparently first brought from the Asian coast to Sweden and is now established as a garden escape on American shores climatically similar to those of its original home.

The Preliterate Islanders

My subject is not anthropological, but man at any stage of culture remains part of the fauna and of the whole ecosystem of whatever region he inhabits. Civilized man, indeed, is far behind in soaking up that basic part of his education.

The members of the established human population at the time of European settlement were woodland Algonkins, divided into numerous so-called tribes but united by being scared of the mighty Iroquois, up the Hudson, to whom

Tulip tree canoe. (Obviously not of a Long Island landscape, this sketch may have been made in the Hudson Valley.)

they are said to have paid an annual token tribute of wampum and smoked clams. The noun Algonkin is alleged to be derived from a Micmac agglutinative phrase meaning "at the place of fish-spearing," which seems particularly appropriate for the aborigines of Long Island. This people wore traditional scalp locks, but did not make the birch-bark canoes with which they have so frequently been pictured. They did manufacture small, rather lopsided wood-skins from single peelings of hickory bark, but they burnt and hewed their larger, more shapely craft out of the soft trunks of the tulip tree which Walt Whitman, unerring in metaphor, described as "the Apollo of the woods—tall and graceful, yet robust and sinewy, inimitable in hang of foliage and throwing-out of limb; as if the beauteous, vital, leafy creature could walk, if it only would."

As reported by Daniel Denton in 1670, the Indians cultivated simple crops and had very wide tolerances as to what constituted game-food. But, most of all, they took their sustenance from the largess of the coastal waters, which their white successors of recent generations have come to prize so extravagantly. Parenthetically, I may say that up to my boyhood cherrystone clams were twenty cents a hundred—instead of, say, fifteen cents apiece. It was still easier to get them for nothing.

We now know from archaeological diggings that the Long Island aborigines had a history of five thousand years, more or less. The number of arrow-points and other celts, literally barrels full, in possession of Long Island farm families means either a great population or very long occupation of the land. We know that the latter is the only tenable hypothesis. The home-made artifacts were of quartz, gathered as beach pebbles, but an occasional flint discloses the seeping process of barter with mainland Indians.

Native density was thin when the white settlers arrived, and subsequent "Indian troubles" were fewer than in most parts of the eastern United States. The Indians seem to have waged but one real battle against the interlopers, this

26

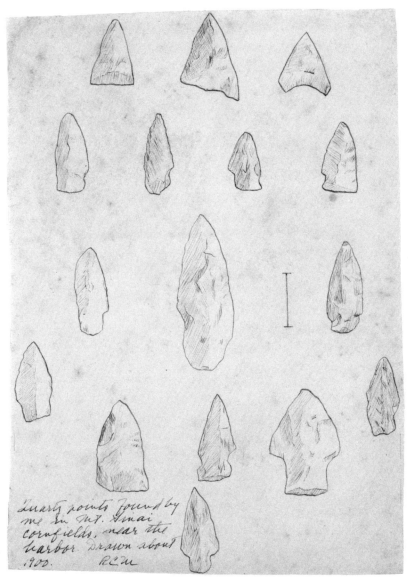

Nonowantuc quartz points, found in cornfields near Mount Sinai Harbor and sketched by the author about 1900. The Nonowantuc Indians were a sub-group of the Setaukets. The line represents one inch. The notched points are prehistoric; triangular points, such as the three in the top row, are of the type in use at the date of European settlement.

27

in 1653 when colonists under command of Captain John Underhill virtually exterminated the Massapequa community between the Hempstead Plains and Islip Township. Denton supplies the clue as to why other wars or skirmishes were unnecessary. His sentiments lack the truculence and vitriolic piety of his Massachusetts contemporary, the Reverend Cotton Mather, but they reflect the same heaven-born attitude that the natives were *intended* to be replaced by a better breed. Denton wrote:

> To say something of the Indians, there is now but few upon the Island, and those few no ways hurtful but rather serviceable to the English, and it is to be admired, how strangely they have decreast by the Hand of God, since the English first setling of those parts; for since my time, where there were six towns, they are reduced to two small Villages, and it hath been generally observed, that where the English come to settle, a Divine Hand makes way for them, by removing or cutting off the Indians either by Wars one with the other, or by some raging mortal Disease.

A map of Long Island bristles like quills of the porcupine with Indian names. In much of the eastern United States some such names are Indian but not indigenous. In the Shawangunk (Algonkian word) Range of the New York mainland we find Lake "Minnewaska," a Siouan term transplanted from the Great Plains. Florida, the "importingest" of all the states, has a "Nokomis," straight out of *Hiawatha*. But on Long Island our geographic names truly smack of the blue mud of eelgrass and scallop flats—Quogue, Amagansett, Wantagh, Mattituck, Yaphank, Matinecock, and Speonk—whether or not the Redskins would fathom our pronunciation. I tend to froth at the mouth when Nonowantuc becomes "Mount Sinai" and when Good Ground, an Indian place name in English, is changed by the real-estate men to "Hampton Bays."

28

Our poet, Walt Whitman, had much to say about Long Island and other eastern geographic names:

> The red aborigines
> Leaving natural breaths, sounds of rain and winds, calls as of
>   birds and animals in the woods, syllabled to us for names.
> What is the fitness, what the strange charm
>   of aboriginal names? Monongahela—
>   it rolls with venison richness upon the palate.

A century ahead of us, he was close enough to the Indians of Paumanok to be caught by the romantic aura of the "noble savage."

> A red squaw came one breakfast-time to the old homestead,
> On her back she carried a bundle of rushes
>   for rush-bottoming chairs,
> Her hair, straight, shiny, coarse, black, profuse,
>   half envelop'd her face,
> Her step was free and elastic,
>   and her voice sounded exquisitely as she spoke.
> My mother look'd in delight and amazement at the stranger,
> She look'd at the freshness of her tall-borne face
>   and full and pliant limbs,
> The more she look'd upon her the more she loved her,
> Never before had she seen such wonderful beauty and purity,
> She made her sit on a bench by the jamb of the fireplace,
>   she cook'd food for her,
> She had no work to give her, but she gave
>   her remembrance and fondness.
> The red squaw staid all the forenoon,
>   and toward the middle of the afternoon she went away,
> O my mother was loth to have her go away,
> All the week she thought of her, she watch'd
>   for her many a month,
> She remember'd her many a winter and many a summer,
> But the red squaw never came nor was heard of there again.

Indians were still among us during my youth; in fact, we yet have the Shinnecock Reservation at Southhampton, and the annual Poospatuck June meeting at Mastic. The

29

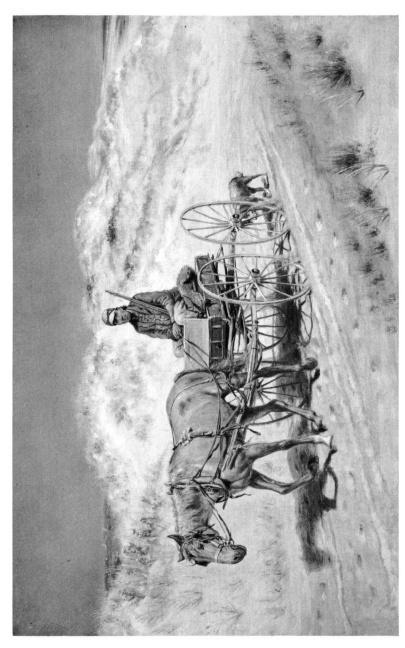

The King of the Montauks. Painted at East Hampton, 1880, by Edward Lamson Henry. (Courtesy of the Detroit Institute of Arts.)

village of the Montauks throve during half the lifetime of my grandfather, even though young Indian men had scattered ever since they became preferred harpooners in the crews of Sag Harbor and Greenport whaleships. My clearest remembrance of an autochthonous Long Islander is of the Shinnecock "Doctor Levi" (Levi Phillips), a herb doctor (pronounced "yarb") who, up to the age of ninety years, carried his satchel of *materia medica* on prodigiously long walks throughout the countryside and prescribed with considerable success for the ills of his red, black, and white fellow men.

An elderly Poospatuck called Martha told me in 1916 of how Doctor Levi had promptly arrested when she was a child a hazardous quinsy throat. The remedy was a gargle prepared by "biling" a certain green plant from her parents' spring-brook. "What do I owe you, Doctor?," asked Martha's mother. "Twenty-five cents, please," replied Doctor Levi, as he resumed his rounds.

The beneficent tradition is, of course, an ancient one on Long Island. Three hundred years before Doctor Levi's ministrations, Daniel Denton wrote "did we know the vertue of all those Plants and Herbs growing there (which time may more discover) many are of opinion, and the Natives do affirm, that there is no disease common to the Countrey, but may be cured without Materials from other Nations."

31

Invaders, Dutch and British

And now for the people who took over, created a civilization, but also ripped the lid off Pandora's box. They had had but a few fleeting forerunners, such as Verrazano, Cabot, and Hudson when, beginning about 1640, they came from New England to Hempstead, East Hampton, Brookhaven (now Setauket), and elsewhere on the Long Island coast or inland beyond the heads of harbors. This, however, leaves out the Dutch, who had been on Manhattan Island since 1623 (they had traded there since 1610) and who soon spread across the East River to found on Long Island Nieuw Amersfoord (Flatlands), Vlackte-Bosch (Flatbush), Nieuw Utrecht, Breuckelen (Brooklyn) and Boswijck (Bushwick). The last was actually first settled by Scandinavians, although in the New Amsterdam realm.

The Hollanders early pushed eastward and by 1650 a dividing line between the Dutch and English colonies was fixed at Oyster Bay. This brought no agreement because both groups claimed the village itself. Final transfer of New Amsterdam from the Dutch to the British crown in 1674 opened up the territories to free movement by the two ethnic groups. Long before the Revolution, Dutch names had become familiar throughout Long Island.

Some of the Dutch chroniclers, such as Adriaen van der Donck (1655) and the Reverend Johannes Megapolensis (1644), antedated publication by Daniel Denton. They wrote chiefly about Manhattan Island and the Hudson valley, rather than Long Island, and their Indians were Mohawk more than Algonkin. Nevertheless, their texts throw much light on our insular field. It is Megapolensis,

32

for instance, who best describes the capacious Indian craft carved from tulip trees and used throughout the coastal area to the south of the range of the canoe-birch.

The central and eastern Long Island settlers, who came from Connecticut or Massachusetts, of course regarded their holdings as part of New England. They sought protection of, and incorporation with, the Connecticut Colony. (At

Well sweep.

least one early deed refers to "the Town of Oyster Bay in New England.") They accepted sovereignty of the Colony of New York only under pressure of necessity. They likewise shared most of the mores of contemporary New England, such as concern with witchcraft. But no witch was ever executed, or even convicted, on Long Island. In the middle seventeenth century, earlier than the great delusion at Salem, several Long Island women were indicted for wickedly practicing the detested art and for secret correspondence with the devil. One was Elizabeth Garlick of East Hampton, a second Mary Hall (and also her husband) of Setauket. The local tribunals were cautious in their judgments, usually putting the suspects under bonds, which were subsequently discharged. There are references, however, to transportation to New England courts "where the proofs necessary to support such accusations are better understood." If the trial was held in Massachusetts, as in the case of Mary Wright of Oyster Bay—in which the charge may not have been witchcraft—there was the added risk of being found guilty of a crime of almost equal enormity, namely of being a Quaker!

Some of the English newcomers were men of cultivation, like the accomplished engineer, Lion Gardiner, who acquired from the Crown but also took pains to purchase fairly from the Indians the island which bears his name and which is still entirely owned by a direct descendant. Numbers of immigrants of this status, such as Floyds, Townsends, Strongs, and Woodhulls, became lords of manors and managed estates extremely well, in part with slave labor. Most *de facto* slavery on Long Island ended long before the Revolution, although its legal abolition in New York State was delayed until 1808.

But the bulk of the immigrants could not be called either good woodsmen or good agriculturists. They were innocent of the sound traditions brought to America by the "Pennsylvania Dutch," who have improved and enriched their farms ever since they first converted them from unbroken forest. The English were chiefly townsmen—laborers, tradesmen, sailors, mechanics, or other small cogs in the current socioeconomic régime. Even those who had come from

34

villages and farms were acquainted only with open-field agriculture and its immemorial practices and could not readily adapt to pioneer conditions. For centuries they and their forefathers had not had to concern themselves with the subjugation of nature, a task that seemed urgent in this New World.

"Bound tree" near Coram, in this instance a white oak which had been lopped when young to form a landmark of ownership.

They had come from a land in which timber, game, fishing, and wild fruits were all entailed in property rights, and in which poaching, and even trespass, were crimes. Here there was no trespass; all was free or nearly so to whomever would take it. The axe preceded the plough. Forest cover was girdled and burnt wholesale. European thrift was promptly forgotten on virgin soil, which was used up with nothing restored, the first-comers neglecting even the simple fertilizing methods of the Indians, such as a moss-bunker (a word taken over from the Dutch) in every corn-

35

hill. The people lived "off the land" and proceeded to wipe out not only proper game but also nearly everything else— all because nature was regarded not as a hospitable environment but as an enemy to be overcome as speedily as possible. They suffered under the "fallacy of the inexhaustible." If woods were laid low, plenty more stood just beyond. The habit of destruction extended not only to the Long Island bear, wolf, bobcat, and rattlesnake—our only venomous serpent—but equally to harmless and beneficent creatures of every sort. Probably the settlers had never heard the puzzled question that the Wampanoags of Massachusetts put to the Reverend John Eliot, who had translated Scripture into their tongue: "Why do Englishmen kill all snakes?" The answer, if there be any, is no doubt in the second chapter of Genesis: "cursed art thou above all cattle, and above every beast of the field; . . . and I will put enmity between thee and the woman, and between thy seed and her seed."

To this day, the influence of symbolic folklore overweighs every reasonable impulse to protect the sad remnant of eleven kinds of nonvenomous Long Island snakes. Of our twelfth species, the rattlesnake, the last known example crawled out of a swamp near Yaphank one spring morning about 1900, and warmed its belly lengthwise on a sun-heated railroad track until a train finished it and its species. Our remaining serpents are harmless, mostly useful from man's point of view, and in several instances notoriously gentle and very beautiful. A milk snake or a ring-necked snake does not struggle or strike in the hand like a captured bird. It remains calm, as though curious and pleased; when released, it moves on without haste or panic. Why such creatures are less popular with us than a chickadee is something that I have never comprehended.

The Long Island colonists, by any fairly applicable

criteria, were, of course, "good folk." They brought with them wives of their own stock. They had the zeal, energy, and vision to become parcel of the making of a nation. But when their personal merits and those of their fellow pioneers are given full credit for the result, it behooves us to remember that success stemmed equally from the fact that they took for nothing and exploited with heedlessness the greatest store of natural treasure that has ever fallen into the hands of mankind. In one sphere, namely the wise use of such resources, the American genius overshot its mark.

Wash day.

Eastern Long Island was, and still is, full of *Mayflower* names, such as "William Brewster" and "John Turner," suggesting the quick progression from Plymouth Rock to Paumanok. And any random roster of original families shows names galore of captains and seamen of the little English fleet which only about half a century earlier had dispelled and scattered to destruction the Invincible Armada.

37

Kitchen of the Mount House, Stony Brook.

Bayles, Conklin, Davis, Dayton, Hallock, Hand, Havens, Hawkins, Hedges, Helme, Homan, Hopkins, Hulse, Jayne, Jones, Ketcham, Miller, Phillips, Platt, Randall, Roe, Satterly, Scudder, Smith, Strong, Terry, Tooker, Warner, Wicks. Children bearing nearly half these names were my fellow pupils in a one-room Long Island school, where Miss Giulietta Hutchinson was an extraordinarily proficient teacher.

During the long decades since settlement, branches of these old families had attained security by way of fishing, whaling, shipping, farming, stock-raising, chandlery, or other trade. They had built comfortable, and in some instances handsome, homes, and had taken on hired hands. And any humane and considerate head of a household or

38

business would refrain, naturally, from forcing his servants or apprentices to partake of too frequent meals of common staples such as venison, wild pigeon, partridge, heath hen, brant, oysters, bay scallops, salmon, sturgeon, or terrapin. Perhaps it was plain thinking and high living that enabled so many old-time Long Islanders to abide here into advanced age. It is all too certain that most forms of life which furnished their early sustenance failed to abide long with them.

Edwin Way Teale, one of the gifted writers on natural history of our time, has told me of his last sighting on Long Island of the plains-inhabiting sandpiper known as the upland plover. The bird was perched atop a wooden sign, and when Teale came within range he read: "A supermarket will be erected here." Exit plover! I suspect that this was the ultimate shock that led my friend to abandon Long Island for an area which may perhaps enjoy a little more borrowed time. I can imagine that our gray poet, dead since 1892, might respond similarly to the compliment of the "Walt Whitman Shopping Center" at Huntington!

Use and Misuse

Much of the northern shore of Long Island, above high sandy or clayey bluffs along the Sound, has today grown up again to dense second-growth forest, which is of coppice type rather than the open stand of primitive times. This cover makes it difficult to realize what happened during the seventeenth century.

Old accounts, as well as old pictures, tell us that by the year 1700 the primeval woods had been mostly cleared off and that the villages, which are now shaded, then loomed

Conditioning the hoofs.

up on treeless fields. We know that this bare state continued throughout the next century as well. During the Revolution the commanders of His Majesty's men-of-war, always on the lookout for provender, would post men with spyglasses in the tops, from which they could look over the bluffs and spot cattle well toward the middle of the island.

Whenever an enemy ship was spied heading westward

into the Sound, local Paul Reveres started riding in relays from the hamlets near Orient and thus spread the alarm to western Long Island. The farmers thereupon hastened to round up their stock and to drive it into concealment in the glacial kettles. Nothing tempting could thereafter be discerned from the mastheads. Behind Smithtown lay a large and deep kettle called "Yorke." It is said that when landing parties questioned the residents, the latter truthfully stated that the cattle had all been sent to Yorke, by which they meant the British seamen to understand "New York."

After the seventeenth-century clearing, much of the poor tillage was succeeded by poor stock-raising. The countryside between villages for the better part of two centuries became indifferent pasture full of half-wild cattle, horses, and sheep. It is of record that the British fleet "requisitioned" (a euphemism) 75 beef cattle and 1,200 sheep from Gardiner's Island alone, a place where the livestock was probably of better quality than throughout most abused Long Island grazing land.

Now the question arises, "Where had the trees gone?" Many, certainly, had been wasted. Others had made homes, outbuildings, fences, grist-mills, churches, schools. And a large proportion had been turned into sea-going craft, for from the earliest days shipbuilding had developed as a major industry. The harbors of Peconic Bay and the Sound coast were ringed with shipyards, which produced vessels for coasting commerce and later for West Indian and European trade and deep-sea whaling. Numerous whaleships of New England registry were actually constructed on Long Island. The brig *Daisy,* of New Bedford, in which I cruised for a year in 1912-1913, was built as late as 1872 on the shores of Setauket harbor, close to my home, and she was the thirty-sixth craft launched from the ways of the shipwright Nehemiah Hand.

New Bedford Whaling Brig *Daisy,* built at Setauket, 1872, by Nehemiah Hand; lost at sea in 1916. Painting by Clifford W. Ashley, owned by the author.

In 1902 I witnessed the launching at Port Jefferson of the 1100-ton schooner *Martha E. Wallace.* This may have rung the knell for home-built merchant-sailers, but sloops and small schooners continued to carry cordwood, coal, oystershells, sawn lumber from Maine, and other bulk produce for several decades more. While I was an undergraduate at Brown I once returned from Providence to Long Island aboard such a schooner. At still earlier dates, when the roads were sandy lanes and the railroad ran only along the midline of the island, most traffic was by water. Up and down the Sound schooners transported passengers as well as cargo, serving meals at unimaginably low prices. They might even convey an orchestra for an elaborate country wedding out from New York and back again, a voyage

42

that could hardly be completed within less than two or three days.

As to consumption of forest by shipbuilding, we have these figures from 1770: a 74-gun frigate required the lumber of 2,220 trees, estimated to be the normal cover of 40 acres of first-growth.

Timber, of course, continued to exist in patches far from the sites of shipyards until later dates. This comprised stands of chestnut—the commonest big tree in Long Island woods until it was destroyed by the blight—as well as oak, locust, spruce, and of planted, rather than indigenous, white pine. These lasted until an era of portable steam sawmills, after the Civil War, took most of the remaining mature trees not sequestered as part of the landscape around homesteads.

Thereafter Long Island was permitted to revert to wood-lots and more extensive tree-cover, mingled with newer and efficient truck-gardening plots and with the admirable potato and cauliflower acreage in such districts as the north

De-ticking man's best friend.

43

fluke. These have been restored to a fertility surpassed nowhere. The principal threat to such land is the well-nigh irresistible demand for more and more homes.

Before leaving the subject of forest destruction, it should be noted that for a century the Long Island Rail Road, employing first wood-burning and later coal-burning locomotives, was the prime starter of innumerable fires. The character of the soil enhanced the damage. The island has ample rainfall to support forest and marsh, but the prevailing sand and glacial pebbles provide such rapid drainage that drought conditions can follow ten summer days without rain. In most adjacent continental districts the soil would preserve a relatively moist state for about twice that long.

Squandering the Fauna

It would be quite incorrect to leave the impression that *all* the natural resources of Long Island were eliminated within any short period. Nature is too long-suffering and tough for that. The wild turkey went out early, along with the wolf and the beaver. The great auk, Labrador duck, heath hen, passenger pigeon, and Eskimo curlew one by one became extinct—as dead as the dodo—but not solely through the destructive propensities of Long Islanders. Ducks, geese, snipes, and plovers remained to be slaughtered for the markets of the metropolis, and they withstood the toll, more or less well, up to a definite point determined by modern firearms, rapid transportation, and an exploding populace.

Waterfowl were shot from "batteries" surrounded by hundreds of floating decoys. Waders were killed from blinds at the edge of salt meadows, as well as on the upland plains. As long as muzzle-loading guns, with their revealing roar and black powder smokescreen, were in vogue, the birds had a chance of sorts, despite the fact that some of the snipes and plovers had to run the same gantlet along the narrow ribbon of their flight-range all the way from Canada to Argentina. But the new guns and motorcars would have quickly sealed their fate if they had not been permanently removed from the game list in 1927. The change has resulted in a modest and, in a few cases, spectacular restoration since that date.

"Forester" and "Cypress" were pseudonyms of Henry William Herbert and William Post Hawes, who in *Sporting Scenes and Sundry Sketches* (1842) have given us a mem-

45

orable description of Long Island shooting in the 1830's. "Human nature is still projectilitarian," the authors observe, which then meant that whatever came within range was fired at. They cite a typical day's bag from the Great South Beach as 20 scoters, one fox, 54 brant, 7 Canada geese, 5 widgeons, 3 oldwives, a cormorant, and a snowy owl—not by any means all edible. But even after 1900 a

Canada geese.   "Laying low for black ducks."

single so-called sportsman once shot in the neighboring bay 350 bluebills or scaup ducks between sunrise and dark.

In my boyhood we were still projectilitarian; the first firearm I used was a double-barreled, muzzle-loading, percussion-cap fowling-piece. Plenty of such shotguns were preserved in old homes, and they were cheap to come by, particularly from elderly women who had lost all their menfolk. We were also inveterate trappers and builders of springpoles and of partridge-hedges with horsehair springes at the openings. I came a little too late to take advantage of the Fulton Market wagoner, who had formerly rolled through the countryside in autumn, paying forty cents apiece for ruffed grouse, even more for duck, fifteen cents for bobwhite, and ten for a rabbit or squirrel. The New York State laws of 1895 were the first to forbid the sale of game "except that taken during the open-season"—a somewhat left-handed compliment to the suppliers! But open seasons were lengthy, baiting ducks and shooting as late or as early as one could see were quite within the law, and bag limits were still of the future.

The news.

Furthermore, the bounty on so-called vermin remained, and it began at twenty-five cents for the smaller predatory mammals. The once abundant skunks had already gone almost out of existence very suddenly. When, about 1894, the Colorado potato beetle reached the Atlantic coast, farmers proceeded to poison the insect with Paris green. The skunk, a prevailingly helpful creature, did its best to assist but failed to discriminate between wholesome and poisoned beetles. It has been extremely scarce on Long Island ever since. Woodchucks and weasels were no longer plentiful enough to be significant, but the opossum was a real revenue-builder. It also passed unconscious judgment on the wisdom of the bounty system. Cash was paid at the town clerk's office upon delivery of the mammalian ears, and tradition among boys was that a captured 'possum, especially a female, should be spared jeopardy of its life unless it wore a salable winter fur. The papery, goblin ears could be snipped off almost bloodlessly, after which the animal was turned loose to breed more 'possums worth a quarter apiece! Muskrats, raccoons, and red foxes made up the principal juvenile fur market. Minks, which I saw now and then, regularly killed some of my Pekin ducks when exceptional tides flooded the salt meadows, but I never succeeded in trapping one.

Stamp of an Insular Culture

The only bearing that the Civil War has on my story is that in its aftermath a large number of young men, discharged from service in the Union Army, left Long Island forever. They may have been the more enterprising members of their generation and communities. The advice of Horace Greeley perhaps never reached their ears, but they set out, nevertheless, to conquer worlds on new frontiers and to carry Long Island names beyond the Rocky Mountains.

I believe that this marked a decline in the status of some of the old Long Island families. Only dry terminal twigs, past fruiting, were left on certain family trees, at least in the home bailiwick. When we speak of a man of good family, we sometimes mean a family that had formerly been better. Such seems to me to have been the status of numbers of older friends and acquaintances of my boyhood. They appeared to have descended not only from, but also below, their ancestors. They no longer wholly matched the books, portraits, diplomas and, in one or two instances, painted wooden escutcheons from a British monarch, which had come to them from forebears of greater distinction. For that matter, the transformation did not necessarily date from a time as recent as the Civil War. Walt Whitman's family appears to have undergone a decrease in intellectual bent and worldly position between the day of his first Long Island ancestor, the Reverend Abijah Whitman, in the seventeenth century, and the outcrop of genius in the poet himself.

In the 1890's the Long Islanders I knew used a vernacular speech different from any I have encountered else-

49

where. It was enriched by a host of archaisms, both of landsmen and of mariners, nearly all traces of which are gone from the conversation of today. The possessive third-

Before the election.

50

person singular pronouns were "hisn" and "hern," which are forms of extremely ancient lineage. "Hern" occurs in the Wyclif English Bible of 1388: "Restore thou to hir alle thingis þat ben hern." Emphatic affirmations or negations were nearly always doubled as "yes-yes" and "no-no." I don't know the English roots of this but its Romance equivalent is familiar in *Oui, oui, Madame,* and *No, no, Señor.* I once chatted with a farmer whom I did not know and heard him say, "Yonder is the house my son built for his wife. We fain would have had them live with us, but they were loth."

Hermaphrodite brigs and fore-and-aft rigs.

In 1932 I lectured at a meeting of the Royal Geographical Society. On the preceding afternoon, a Sunday, the President, Admiral Sir William Goodenough, and his Lady gave a tea for Mrs. Murphy and me in their Surrey home and garden. "This place," said our host, "as you have seen at the gate, is called 'Parson's Pightle.' The latter is a virtually obsolete term and no doubt obscure to you."

"It may be obsolete in England," I replied, "but I grew up just across the cove from Strong's Pightle, and the pightles of Davis, Sylvester, Warner and others were likewise familiar to me."

"My word!" exclaimed the Admiral.

Pightle, meaning a croft or close, dates as I have since learned from the early thirteenth century. Having our host at a disadvantage, I went on: "Here at the hub of the English-speaking world, Sir William, speech evolves and transforms with embarrassing rapidity, but on Long Island we still converse as Elizabethans, with overtones from Chaucer."

This time the Admiral said, "My God!"

Most of us have perhaps heard more about the conservative speech of the Appalachian mountaineers, with whom the old-time Long Islanders shared other traits as well. For example, my boyhood neighbors were mostly natural and contented feudists. They rarely resorted to the reputed violence of the Kentuckians but they were quick with both threats and applications of what they called the "lor." Courts of justices of the peace conducted a brisk business. There was honorable precedent for this, as shown by a communication from the residents of the Town of Oyster Bay to those of the Town of Huntington on the fifth of July, 1669:

> Friends and neighbors. We once more desire you, in a loving friendly way, to forbear mowing our neck of meadow, which you have presumptuously mowed these several years; and if after so many friendly warnings you

Spritsail rigs and cutters: harbor activities.

53

will not forbear, you will force us, friends and neighbors, to seek our remedy in law; but resting your friends and neighbors.

MATHEW HARVEY, *Town Clerk.*

I doubt whether the most tetchy acquaintances of my early days would admit that they either were or had enemies. Far from it; they merely disliked their friends. They worked together nonetheless cheerfully and their aversion was only transitory; it terminated at every funeral. I recall these neighbors in a period centering around the Spanish-American War as falling into three categories: (1) churchgoers (mainly Congregational in my village and roundabout); (2) Bible-reading nonchurchgoers (which included the bulk of the seafaring men); (3) non-Bible-reading nonchurchgoers (possibly in the majority). The last group had undeniably ancient standing; one preacher wrote, even earlier than the date of Daniel Denton, that our people are of two kinds: (1) Indians; (2) Christians so-called.

I knew several men and women of old age who had never made the journey of sixty miles to New York City. But the greatest contrast between that long-ago period and now lay in the blissful lack of haste. The neighbor nearest my parents' home, Captain Adolphus Davis, drove an ox named Jacob, hitched to a wagon. He never lacked time for a shopping journey of three miles over the hills to Port Jefferson and home again, behind a beast that could not be hurried beyond a walk.

Population Ferment

It is remarkable that a varied, beautiful, and fertile insular tract, lying at the threshold of the greatest urban center in the nation, could remain rural, with discrete and self-sufficient towns and villages, for 250 years following settlement. Even the replacement, after 1844, of the old post-road and ferry route to Boston by a railroad to Greenport made little change in the stamp of Long Islanders.

The island then retained a wide, indeed an international, reputation for sport-shooting. This was probably first established by British officers during the Revolution, when after the Battle of Long Island the enemy took control of most of the area. But its discovery as a playground *par excellence* for city-dwellers came much later, and dense or even congested population is a phenomenon of the present century.

In 1911, when I was graduated from college, my Township of Brookhaven had 16,000 inhabitants. By 1962, 124,000 more had been added. Suffolk County, which had 35,000 souls in 1812 and 77,582 in 1900, is now pushing hard toward three-quarters of a million. The population of the Township of Oyster Bay, in Nassau County, increased by 141 per cent in the five years following 1950. Today only six states in the Union have as many people as reside on Long Island. Not even California, where for some reason they seem to rejoice in it, is growing so fast.

With the whole burst of change so recent, it is inevitable that much of its sweep should have come under my eye. I saw the village blacksmith, Mayhew Woodhull—two great Long Island names, by the way—run out of gainful occupation until he had shifted over to become a catcher and pur-

55

veyor of eels. Wilson Brothers' Sail-Loft, founded in 1837 at Port Jefferson, where John Paul Jones had fitted out, sewed the courses of countless square-riggers, and the canvases of the schooner *America* before she crossed the Atlantic to lift the Cup that has stayed here. But not even the later heyday of yachts could save this firm from becoming the Wilson Awning Co., and finally fading out.

Stencil on the canvas of the schooner yacht *America*, 1851. (Courtesy of Mr. F. Henry Berlin.)

New Lamps for Old. Whither?

What of the future? We cannot answer with assurance. Most present and prospective hazards derive from population pressure, which seems to be piling up its head of steam. If the problem is already vaguely troubling us, how soon may it turn imminent and then desperate?

Vast works have been carried out on Long Island for the quick transport of hordes of human beings to their goals, yet we usually seem about the same distance behind. The drive and competence of Robert Moses have gone far to provide, for the moment, greatest good for the greatest number. But we know that there is a point at which numbers alone destroy the greatest good. Parks and parkways and public resorts we have, but relatively little provision is in view for seizing and saving the remaining wild land and the ultimate sources of life's indispensable mineral, which is water. In fact, at the beginning of every project for the benefit of generations not yet born, opponents seem able to shout down and vote down the backers. Public opinion has by no means kept abreast of long-term requirements. As General Omar Bradley has said:

> Year after year our scenic treasures are being plundered by what we call our advancing civilization. If we are not careful, we shall leave our children a legacy of billion-dollar roads leading nowhere except to other congested places like those they left behind. We are building ourselves an asphalt treadmill and allowing the green areas of our nation to disappear.

Bulldozers, marine dredges in search of gravel, and the point of view of the Army Engineers seem enormously popular on Long Island, although all three have been many times guilty of destructive trends in land use. Ponds and natural catchment basins are still being filled up. The subsequent attempts to control floods are by shunting water through pipes to the sea instead of restoring it to the aquifers from where it would continue to work for us. The slogan "Wetlands are not waste lands" has thus far failed to take hold. A blind spot appears in many civic and political eyes with reference to the still lonely and relatively unravaged Great South Beach. Here is one of the finest ocean beaches in the world, and mostly within two-score miles of the metropolis. It seems crystal clear that its one destiny should be to become the property of the federal government, before time runs out, as the "Fire Island National Seashore."

Recent and dreadful tampering with the Long Island environment includes filling the water table with detergents that cannot be broken down by bacteria, and spraying the face of the earth indiscriminately with chlorinated hydrocarbons. The target of the latter is usually a single species of real or alleged insect pest, toward the hoped-for elimination of which every other susceptible organism must be callously sacrificed. And it seems to be taken for granted that native vegetation should be ruthlessly scraped off sites preparing for public buildings. Subsequently a basal planting is perhaps provided of European yew, which bears the appropriate scientific name *Taxus taxus*.

There is much that is ominous and awry. We amplify and beautify the centers of congregation for fast redoubling human populations without giving equivalent thought to the living space of man's fellow creatures, or to the soil and ground water which are limiting "trace elements" of existence. Shall those who come a few generations after us be able to look anywhere on a luxuriant, vital, and fecund, yet uncrowded world, such as Walt Whitman pictures so stirringly in his myriad verses? Admittedly, we can never dupli-

Crane Neck Point from the inlet of Stony Brook Harbor.

cate his feat of July 25, 1881, when he rejoiced in "a swim and a naked ramble" on the beach of Far Rockaway. Since then a million eyes and a police department have moved in!

Sometimes it is problematical whether Walt's descriptive passages relate to his beloved Paumanok, to the continental mainland, or to a fusion of the two. This is a privilege of the arts. His contemporary, the painter William Sidney Mount of Stony Brook, the recognition and vogue of whom arrived almost as tardily as those of the bard, once spread on canvas a "Long Island landscape with imaginary mountains." We can hardly expect old Walt to be more inhibited by reality or less rebelliously creative.

> Press close bare-bosom'd night—press close magnetic nourishing
>   night!
> Night of south winds—night of the large few stars!
> Still nodding night—mad naked summer night.
> Smile O voluptuous cool-breath'd earth!
> Earth of the slumbering and liquid trees!
> Earth of departed sunset—earth of the mountains misty-topt!
> Earth of the vitreous pour of the full moon just tinged with blue!
> Earth of shine and dark, mottling the tide of the river!
> Earth of the limpid gray of clouds brighter and clearer for my
>   sake!
> Far-swooping elbow'd earth—rich apple-blossom'd earth!
> Smile, for your lover comes.

References

CONARD, H. S. 1935. "The Plant Associations of Central Long Island," *American Midland Naturalist* **16:** 433-516.

DENTON, DANIEL. 1670. *A Brief Description of New-York: Formerly Called New-Netherlands* (London, printed for John Hancock). (The author has two reprints: 1845, in Gowan's *Bibliotheca Americana,* which is especially valuable for its notes; and the 1937 facsimile, Columbia University Press, New York.)

DONCK, ADRIAEN VAN DER. 1655. *Beschryvinge van Nieuvv-Nederlant* (t'Aemsteldam, Evert Nieuwenhof). English translation by Jeremiah Johnson: "Description of the New Netherlands . . . ," *New-York Historical Society Collections,* ser. 2, **1** (1841): 125-242.

DWIGHT, TIMOTHY. 1821-1822. *Travels in New-England and New-York* (4 v., New Haven). (The record of Long Island travels is in vol. 3, 1822.)

"FORESTER, FRANK" (= Henry William Herbert), ed. 1842. *Sporting Scenes and Sundry Sketches; Being the Miscellaneous Writings of J. Cypress, Jr. (= William Post Hawes)* (2 v., New York, Gould Banks & Co.).

FULLER, MYRON L. 1914. "The Geology of Long Island, New York," *U. S. Geological Survey Professional Paper* **82** (Washington).

HELME, ARTHUR H. 1927. "Reminiscences of a Long Island Naturalist," *Brooklyn Museum Quarterly* **14:** 98-105.

MEGAPOLENSIS, JOHANN [ca. 1644?] Een Kort Ontwerp van de Mahakvase Indianen . . . Aldus beschreven ende nu kortelijck den 26 Augusti 1644 opgesonden uty nieuwe Neder-lant . . . Mitsgaders een kort verhaal van het leven ende Statuere der Staponjers, in Brasiel. t'Alskmaer, Ysenbrant Jansz. Van Houten. English translation: "A Short Account of the Maquaas Indians in New Netherland . . . written in the year 1644" in Ebenezer Hazard, *Historical Collections* (Philadelphia, 1792) **1:** pp. 517-526.

MURPHY, R. C. 1899-1939. Notebooks on the geography and natural history of Long Island in the author's library.

————1940-1962. Bound manuscripts on the geography and natural history of Long Island in the Library of the American Philosophical Society, Philadelphia.

————1950. *August on Fire Island Beach,* Science Guide No. 134 (New York, American Museum of Natural History).

———— 1962. "Mitchill's 'Birds of Plandome,' Long Island," *Proceedings of the American Philosophical Society* **106:** 48-52.

STRONG, KATE W. 1940-1962. True Tales from the Early Days of Long Island. Pamphlets numbered "Series 1-24," reprinted from the *Long Island Forum.* Nos. 1-3 published at Bay Shore; 4-24 at Amityville.

SVENSON, H. K. 1936. "The Early Vegetation of Long Island," *Brooklyn Botanic Garden Record* **25:** 207-227.

TAYLOR, NORMAN. 1927. "The Climate of Long Island," *Cornell University Agricultural Experiment Station Bulletin* **458:** 1-20.

THOMPSON, BENJAMIN F. 1839. *History of Long Island* (New York, E. French). (This work has been revised, enlarged, and published in several subsequent editions.)

TRANSEAU, E. N. 1913. "The Vegetation of Cold Spring Harbor, Long Island. The Littoral Succession," *Plant World* **16:** 189-209.

VEATCH, A. C., C. S. SLICHTER, ISAIAH BOWMAN, W. O. CROSBY, and R. E. HORTON. 1906. "Underground Water Resources of Long Island, New York," *U. S. Geological Survey Professional Paper* **44** (Washington).

WHITMAN, WALT. 1949. *The Poetry and Prose of Walt Whitman,* edited by Louis Untermeyer (New York, Simon and Schuster).

Index

Adirondack, *see* mountains
Alewife, 19
Algonkin, significance of term; tribute to Iroquois, 25
Alps, *see* mountains
Altai, *see* mountains
America, schooner yacht; Cup, 56
American Philosophical Society, vii, 3
Appalachian, *see* mountains
Arbutus, trailing, 22
Archaisms in L. I. speech, 50
Army Engineers, 57
Arrow-points, 27

Bartram, John and William, 10
Batteries, in wild-fowl shooting, 45
Battle of L. I., 55
Bay, Chesapeake, Delaware, New York, 24; Peconic Bay shipyards, 41
Beach plum, praised by Henry Hudson, 14
Bear, 36
Bearberry, 22
Beaver, dams, teeth, trapping, 18; extirpation, 45
Beetle, potato, spread to Atlantic coast, 48
Blight, chestnut, 43
Blinds, in shorebird shooting, 45
Bobcat, 36
Bog, on L. I., 17
Boulders, glacial erratic, 4
Bounty on predatory animals, 48
Bradley, Omar, on destruction of scenery, 57
Brant, 39
British Navy, requisition of provender during Revolution, 41
British officers, sport shooting during Revolution, 55
Brookhaven (Setauket), 32; Township, population growth, 55
Brooklyn, settled by Dutch, 32
Brooks, dried up after felling of forest, 17

Bulldozers, destructive effects, 58
Bushwick, settled by Scandinavians, 32

Cabot, preceded settlement, 32
Cactus, *see* prickly pear
Canadian Life Zone, representative organisms of, on L. I., 23
Canoe, Indian woodskin and dugout, 26, 33
Cape Cod, 16, 23, 24
Cape Hatteras, 24
Cardinal (bird), 22; flower, 17
Carolina Capes, 23
Cattail, displacement by Old World reed, 18
Caucasus, *see* mountains
Chestnut, commonest large tree until its disappearance, 43
Chickadee, Carolina, 24
Chlorinated hydrocarbons, sprayed as pesticide, 58
Churchgoers and nonchurchgoers, 54
Cicada, 17-year, 3
Civil War, departure from L. I. of discharged soldiers; related decline of certain old families, 49
Clams, former market price, 26
Clay, Cretaceous, 5
Climate, of L. I., unlikeness at two ends; influence of ocean and continent, 9; rhythms, 10; "year without a summer," 11; contrast with Mediterranean; seasonal amplitudes, 22; delicacy of plant adjustment to, 24
Coastal plain, Atlantic, 1; flora of, 23-24
Cod, 23
Cold Spring Harbor, 17
Coney Island, Henry Hudson landed, 14
Connecticut Colony, incorporation with sought by L. I. settlers from New England, 23

63

...ng Island from the air, 1962. Largest island off our Atlantic ...ong Island has an area of 1,682 square miles, a width of 23 ...shore lines totalling about 500 miles. The coast has intimate ...his air photograph, which reveals also indications of glacial ...lant cover, and the hundred or more square miles of salt-...otograph used by courtesy of Lockwood, Kessler and Bartlett, ...ting Engineers, Syosset, N. Y.)

All
seaboard,
miles, and
clarity in
moraines,
marsh. (
Inc., Con

67

All Long Island from the air, 1962. Largest island off our Atlantic seaboard, Long Island has an area of 1,682 square miles, a width of 23 miles, and shore lines totalling about 500 miles. The coast has intimate clarity in this air photograph, which reveals also indications of glacial moraines, plant cover, and the hundred or more square miles of salt-marsh. (Photograph used by courtesy of Lockwood, Kessler and Bartlett, Inc., Consulting Engineers, Syosset, N. Y.)